Mathematical Olympiac

for

Elementary School 3

My First Book of Mathematical Olympiads – *Third Grade*

(Workbook)

Educational Collection *Magna-Scientia*

My First Book of Mathematical Olympiads

Mathematical Olympiads *for* Elementary School

3

Third Grade

(Workbook)

Michael Angel C. G., Editor

Preface

The *Mathematical Olympiads for Elementary School* are open mathematical Olympiads for students from 1st to 4th grade of elementary school, and they have been held every year in the city of Moscow since 1996, their first editions taking place in the facilities of the Moscow State University - Maly Mekhmat. Although initially these Olympiads were conceived for students of a study circle of elementary school, then it was extended to students in general since 2005. Being the Technological University of Russia – MIREA its main headquarters today. Likewise, these Olympiads consist of two rounds, a qualifying round and a final round, both consisting of a written exam. The problems included in this book correspond to the final round of these Olympiads for the 3rd grade of elementary school.

In this workbook has been compiled all the Olympiads held during the years 2011-2020 and is especially aimed at schoolchildren between 8 and 9 years old, with the aim that the students interested either in preparing for a math competition or simply in practicing entertaining problems to improve their math skills, challenge themselves to solve these interesting problems (recommended even to elementary school children in upper grades with little or no experience in Math Olympiads and who require comprehensive preparation before a competition); or it could even be used for a self-evaluation in this competition, trying the student to solve the greatest number of problems in each exam in a maximum time of 1.5 hours. It can also be useful for teachers, parents, and math study circles. The book has been carefully crafted so that the student can work on the same book without the need for additional sheets, what will allow the student to have an orderly record of the problems already solved.

Each exam includes a set of 8 problems from different school math topics. To be able to face these problems successfully, no greater knowledge is required than that covered in the school curriculum; however, many of these problems require an ingenious approach to be tackled successfully. Students are encouraged to keep trying to solve each problem as a personal challenge, as many times as necessary; and to parents who continue to support their children in their disciplined preparation. Once an answer is obtained, it can be checked against the answers given at the end of the book.

Sincerely,

The editor

Contents

Problems

Olympiad 2011

(XV Olympiad for Elementary School)

Problem 1. In a family there are three fathers, three children, a great-grandfather and a great-grandson. How many people are there in all?

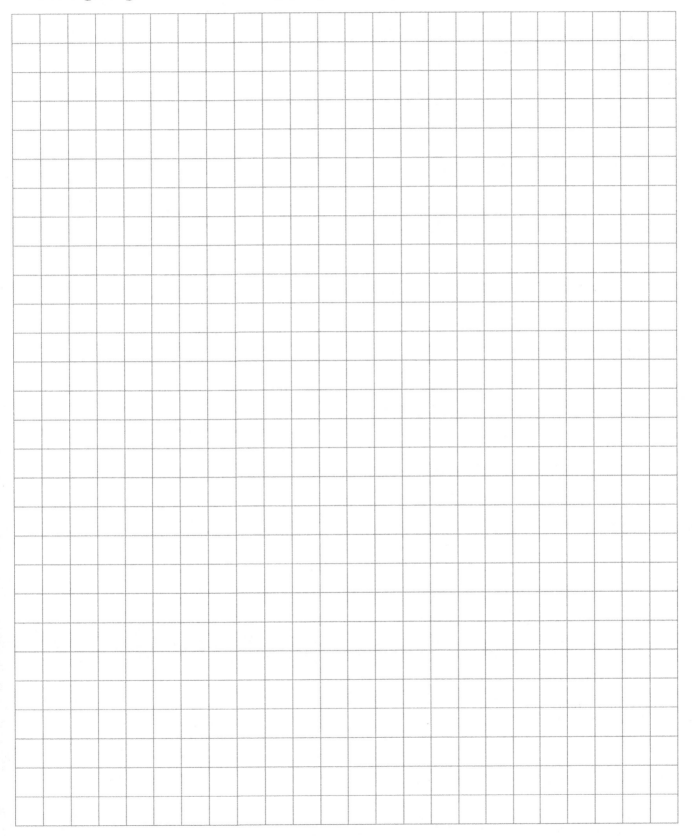

Problem 2. Definition 1. "A glove is a wool product in which the fingers are warm".
Definition 2. "A shoe is something that is put on the foot and has laces".
Definition 3. "A sock is a wool product that is worn on the foot".
Definition 4. "A SHOEGLOVE is both a glove and a shoe".
Is the SHOEGLOVE a sock?

Problem 3. Cut the snowflake shown in the figure into 5 pieces with a straight cut.

Problem 4. Kopatych weighs more than Losyash. Hedgehog and Losyash together weigh more than Nyusha and Kopatych together. But Kopatych and Losyash together weigh the same as Hedgehog and Nyusha together. Who weighs more and who weighs less?

Problem 5. Dima has two large coils of rope: white and black. He cuts 10 cm pieces from the coils and ties the three pieces into a 30 cm ring. How many different rings can he get?

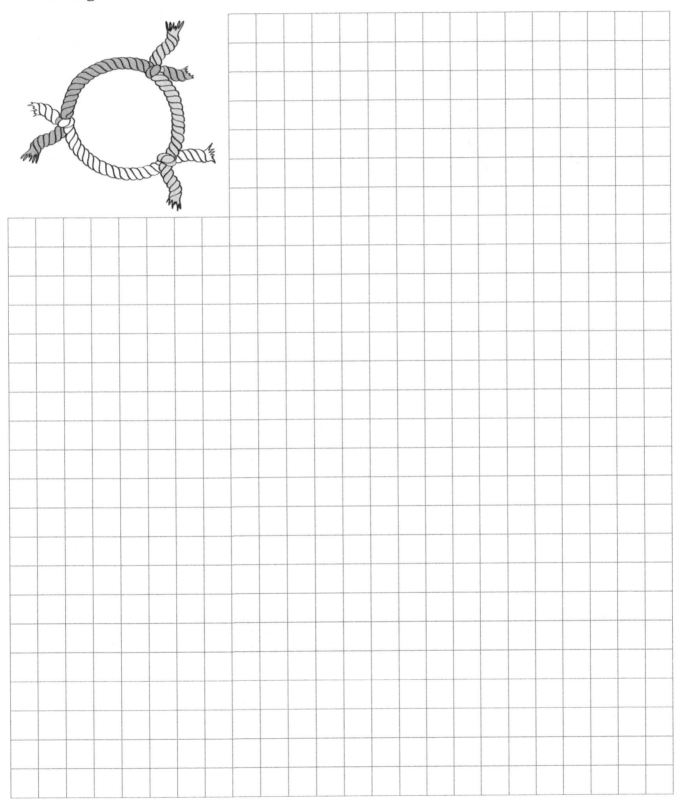

Problem 6. Ranger Stepanych roams the boundaries of his place in 4 hours. While the Ranger Mikhalych walks the boundaries of his place in 6 hours. When Mikhalych retired, his place was incorporated into the Stepanych place and now Stepanych spends 8 hours roaming the boundaries of the combined places. How long did it take him to travel the shared boundary of these places, if the speeds of the rangers are the same?

Problem 7. Vanya has 8 dominoes (see picture). She wants to arrange them in the shape of a 4 × 4 cell square so that the sum of dots in all the rows and all the columns of the square is the same. A) What should this sum be equal to? B) How does Vanya need to arrange the dominoes?

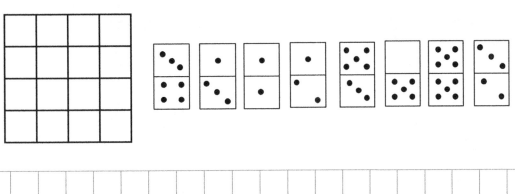

Problem 8. John Silver hid a treasure of gold and silver on three islands: Green Island, Amber Island, and Rocky Island. In one he hid the gold, in another he hid the silver, and in the remaining island he hid nothing. In the bay of each island he posted signs. On the green island: "Gold on the rocky island". On the amber island: "There is no gold or silver here". And on the rocky island: "There is no silver either on the green island or on the amber island". Where is there definitely nothing if all the signs are not telling the truth?

Olympiad 2012

(XVI Olympiad for Elementary School)

Problem 1. Between some digits, put an equal sign and an arithmetic sign to get the correct equality:

$$2\ 0\ 0\ 0\ 2\ 0\ 1\ 2\ 1\ 2$$

Problem 2. Znayka photographed reflections of the apples in the mirror. If Dunno replaced one picture with another. What apple is Znayka missing the photo of?

Problem 3. Cut the checkered shape of the figure below into four equal pieces, each with a marked cell.

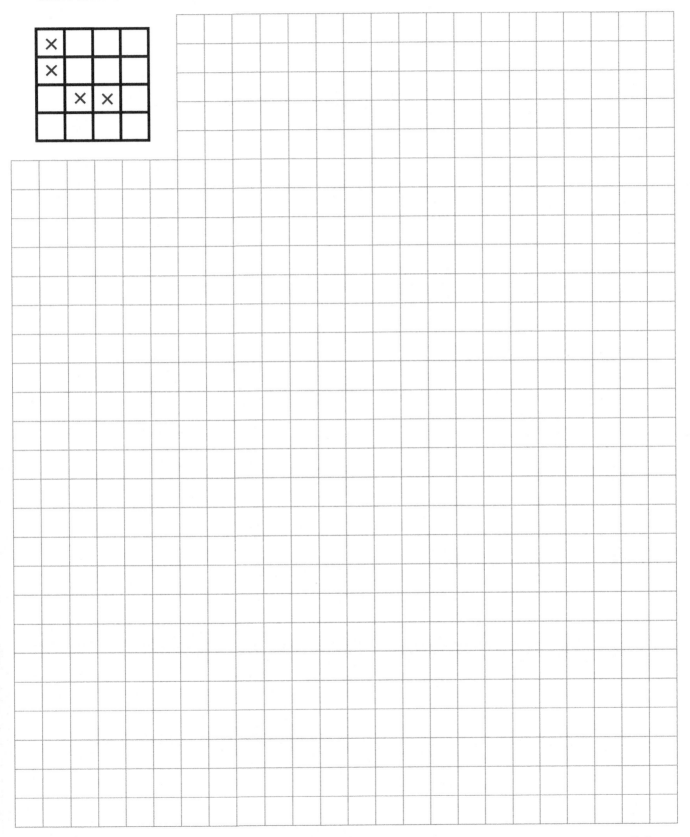

Problem 4. In the table below, arrange the numbers 1 through 5 so that each column and each row, as well as each highlighted shape, contains all five numbers.

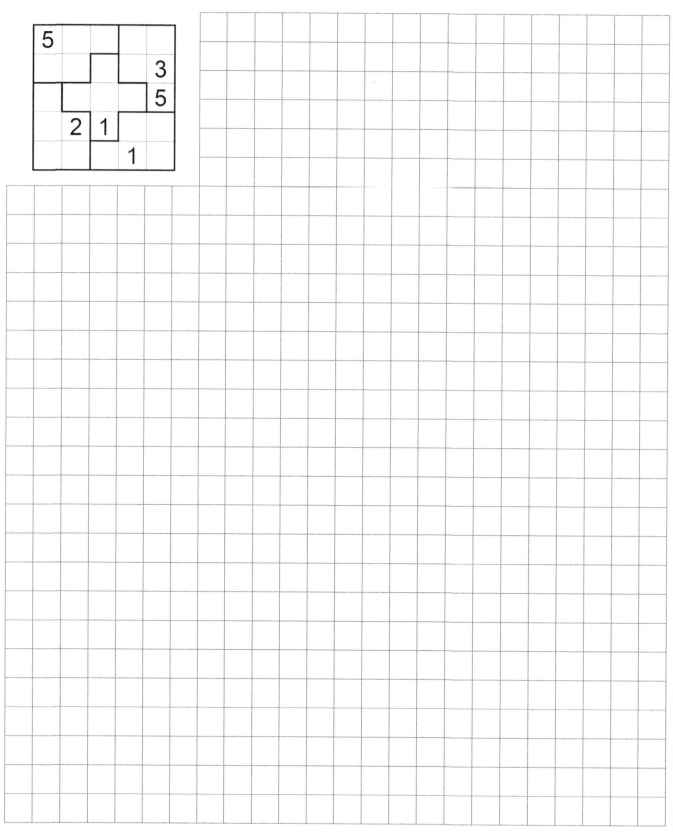

Problem 5. Uncle Fyodor's pace is three times that of Matroskin. First, Matroskin walked a straight path, and then Fyodor, starting from the same place as Matroskin. Following Matroskin's trail, Fyodor erases this trail. So Sharik counted 17 footprints of Matroskin. How many footprints of Fyodor were on the path?

Problem 6. Winnie the Pooh has 11 large jars of honey and 10 small ones. A store sells boxes in which can be packed 5 large, 9 small, or 4 large and 3 small jars. How many boxes does Winnie have to buy to pack all of his jars of honey? (If he wants to buy as few boxes as possible).

Problem 7. Several Indians and some pale-faced people formed a circle. It is customary for them to lie to their own people and to tell the truth to people with a different skin color. Each turned to his neighbor on the right and said a phrase. If there were 8 phrases "You are an Indian" and 9 – "You are pale-faced". How many Indians and how many pale-faced are there?

Problem 8. Sasha has 2 gold coins, 3 silver and 4 bronze; and one of them is false. If the false coin is silver, then it is lighter than a real silver; and if the gold or bronze coin is false, then it is heavier than the real gold or bronze coin, respectively. Find the false coin after two weighings on a two-plate scale. (Note. Coins made of different metals may weigh differently, but real coins made of the same metal weigh the same.)

Olympiad 2013

(XVII Olympiad for Elementary School)

Problem 1. In the next sum, different letters represent different digits. It turns out that OLIM + PI + ADA = 2013. Indicate which numbers could be in place of the letters.

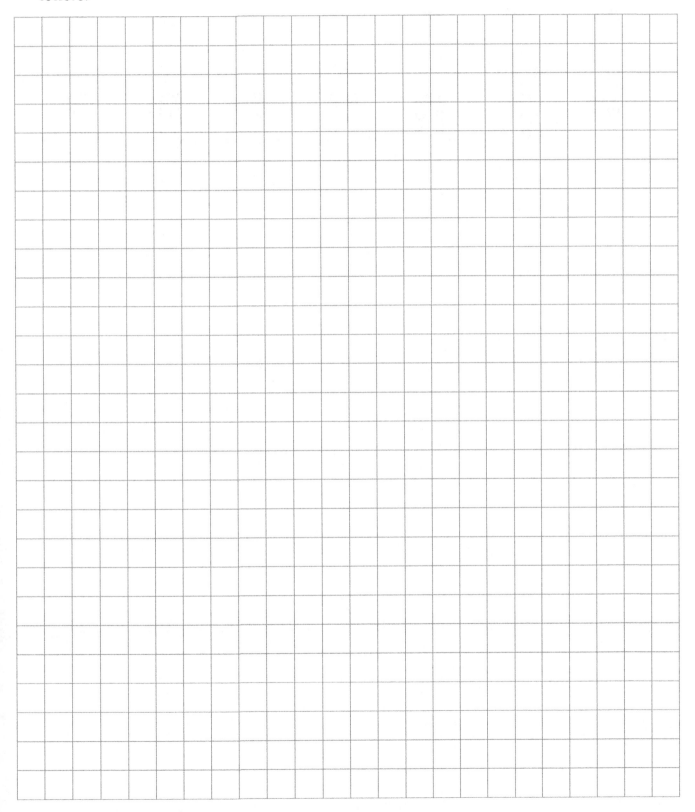

Problem 2. My brother always wants to dress differently from me. Therefore, his clothes and shoes are completely different from mine. Determine my brother's name.

ME   PASHA   YASHA   MAXIM   TOLYA   KOLYA   OLEG

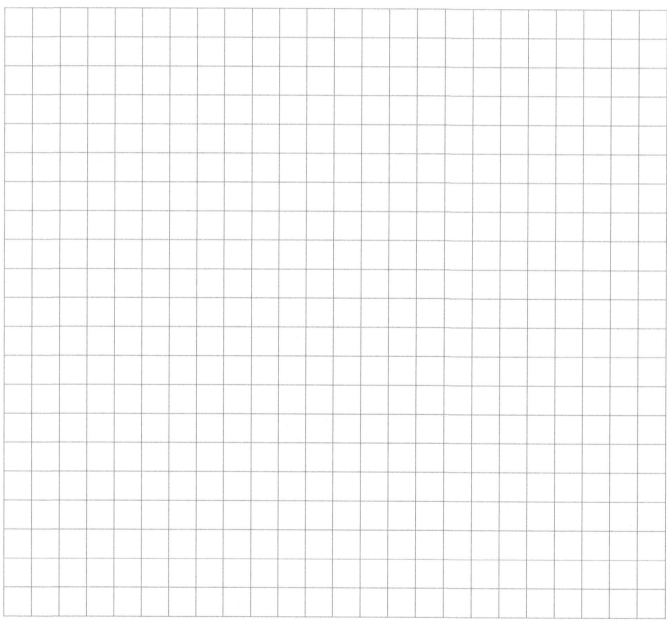

Problem 3. In 3rd grade, 6 people eat ice cream every day, 8 people eat ice cream every other day, and the rest eat no ice cream at all. Yesterday 12 students in this class ate ice cream. How many students will eat ice cream today?

Problem 4. In the figure made of matches, five triangles can be counted: four small and one large. Move two matches so that exactly four triangles are visible. There should be no additional matches.

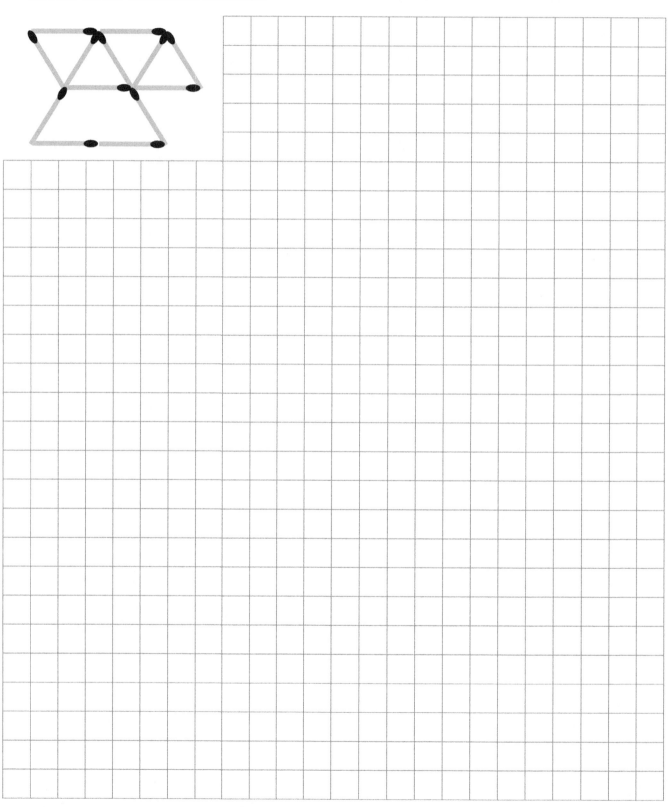

Problem 5. The young physicist Ilya has two identical elastic bands. Likewise, he marked the midpoint of each of them and hung weights on their ends so that one elastic band became twice as long as the other. Ilya measured how far one mark is below the other. How many times is this distance less than the length of the longest elastic band?

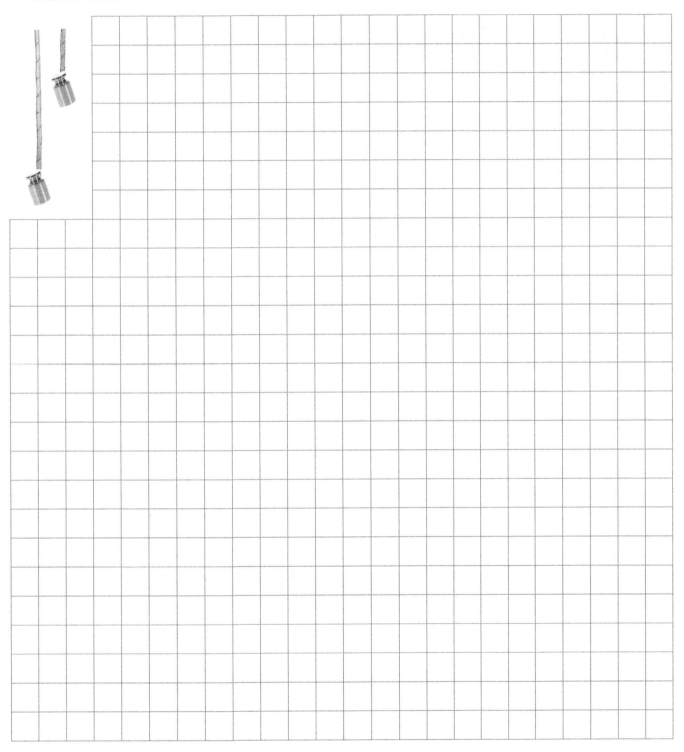

Problem 6. Anya, Borya, Vasya and Galya decided to eat a bar of chocolate. But it fell to the floor, and when they picked it up, it turned out that it was broken into five pieces, as shown in the figure. Borya ate the largest piece. Galya and Anya ate the same amount of chocolate, but Galya ate two pieces and Anya one. Vasya ate the rest. What piece of chocolate did Vasya get?

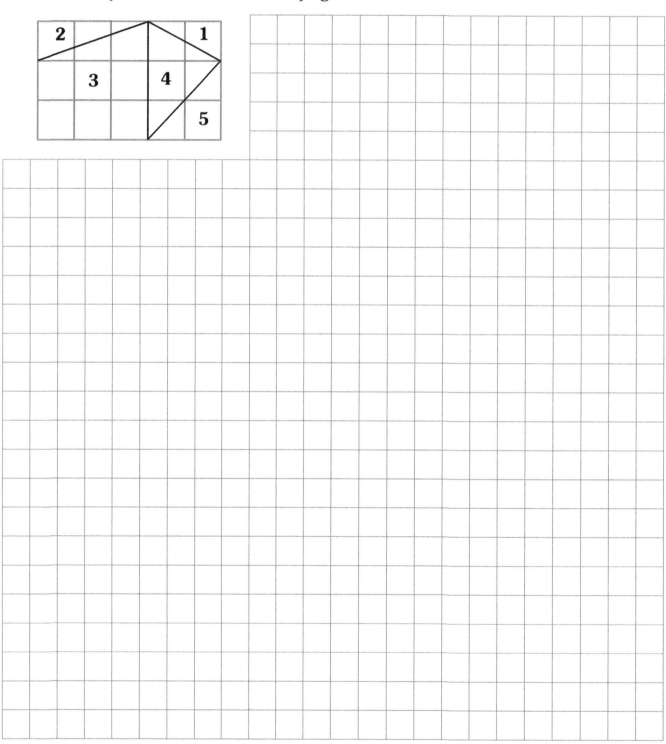

Problem 7. A tinsmith makes letter signs. He makes the same letters at the same time, and different letters, perhaps at different times. If he spent 50 minutes on two signs "DOM MODA" and "DINO", and made the sign "DOMINADO" in 35 minutes. How long does it take to make the sign "ANIDO"?

Problem 8. One day, three friends were talking at a party. Gloria said, "I always speak less than six words". Alex said, "All sentences of more than six words are false". Marty added worriedly, "At least one of us is lying right now". Determine who lied and who told the truth.

Olympiad 2014

(XVIII Olympiad for Elementary School)

Problem 1. Petya, Vasya, Olya and Masha line up for ice cream. It is known that, the girls are not together, Vasya is right behind Olya and Masha is right behind Petya. What is the order in the line?

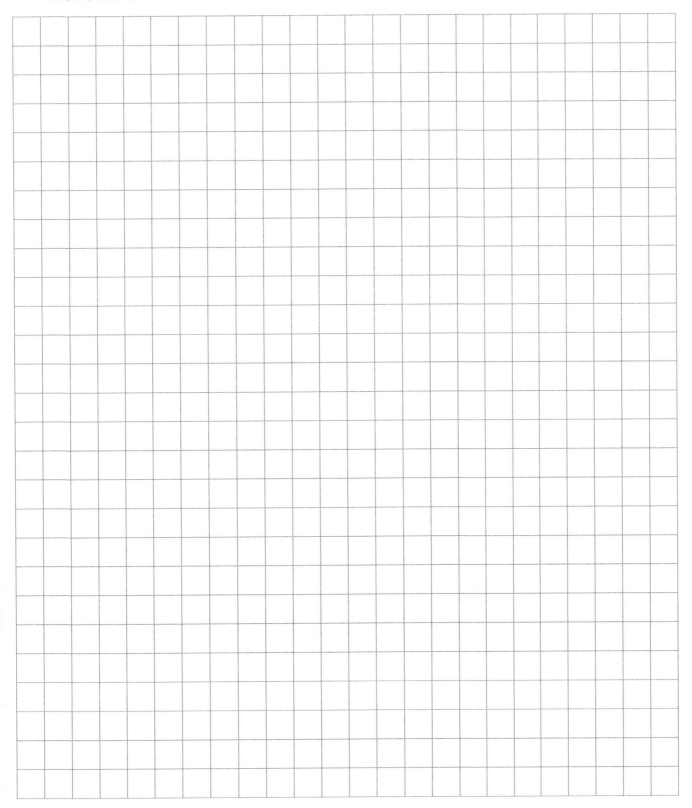

Problem 2. Shpuntik was driving a car and saw a kilometer post, on which the number of kilometers was written with a two-digit number with different digits. He drove a little further and saw a kilometer post with the same digits as before, but written in a different order. What is the smallest distance between these posts?

Problem 3. Eight gears are meshed with each other and rotate. The lower left gear rotates clockwise, as shown in the figure. In which direction does the lower right gear rotate?

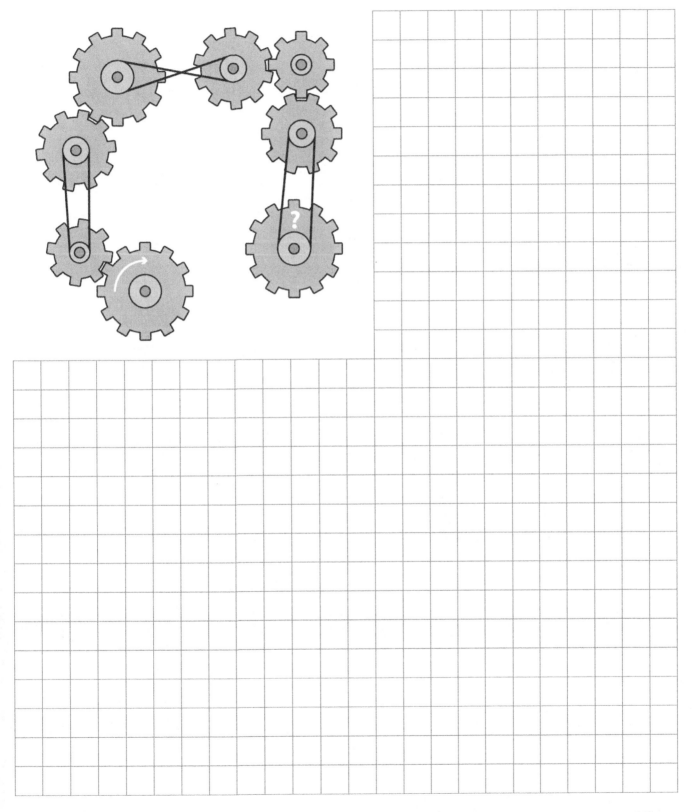

Problem 4. For the New Year, Santa Claus brought candy for the 3rd graders. If he gives each girl 3 candies and each boy 2, then he will need one more candy. And if he gives each boy 3 candies and each girl gives 2 candies, then he has two candies left over. What is more in the class: boys or girls, and how much more?

Problem 5. Malish and Carlson each had an identical triangular pyramid made of paper. They cut them out in two different ways. If the figure A shows what Malish got. How will Carlson's cutout be painted (see figure B)?

(A)   (B)

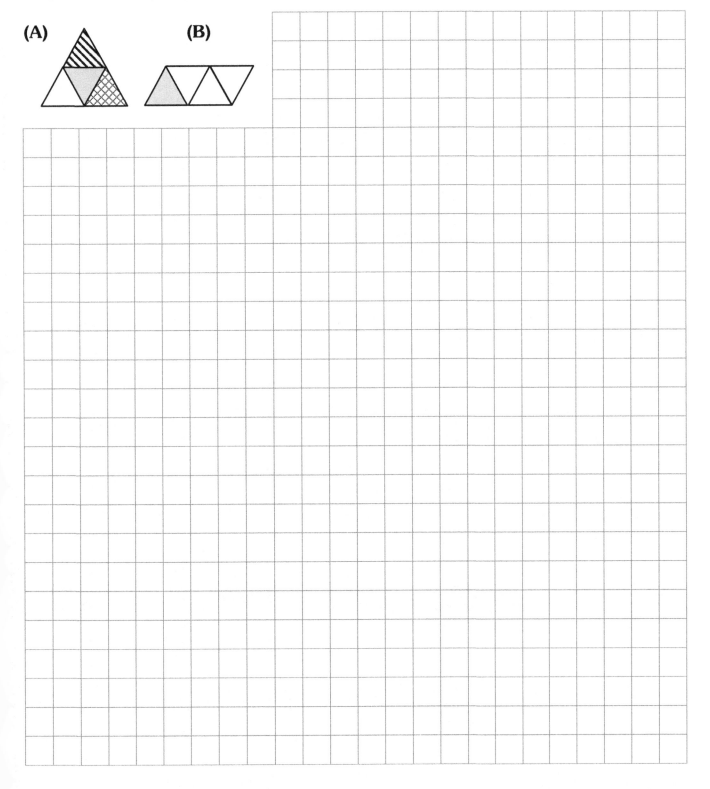

Problem 6. Examples of additions were written on a blackboard. Little Johnny replaced different digits with different letters. It turned out that, $O + N + E + F + O + R + O + N + E = 32$, and $S + I + X + F + O + R + S + I + X = 38$. What is the next addition equal to: $O + N + E + F + O + R + S + I + X$?

Problem 7. Popeye "The Sailor" only eats spinach, and exactly once a day: either for breakfast, for lunch or for dinner. It is known that if Popeye has lunch one day, the next day he will definitely not have breakfast. If he has dinner one day, the next day he will definitely not have breakfast or lunch. If he has only had dinner twice in the last 2 weeks. At which meal of the day did Popeye eat his spinach yesterday?

Problem 8. Brothers Avoska and Neboska only lie on their birthdays. Other days, they just tell the truth. Avoska once said: "Today is April 1. Tomorrow is your birthday". Neboska replied, "Today is your birthday. Tomorrow is April 1". When was Avoska born?

Olympiad 2015

(XIX Olympiad for Elementary School)

Problem 1. How many three-digit numbers that have the sum of their digits equal to 3 are there?

Problem 2. Kopatych went into hibernation on November 15. But every eighth day funny hares woke him up and he came out of his den. If Kopatych came out of his den ten times. On what date did he leave his den for the last time?

Problem 3. There are 16 boxes on a table. Each one contains an apple or an orange, or the box is empty. It is known that there is an apple and an orange in each row (vertical and horizontal line). Labels are placed around the boxes, indicating which fruit is closest to the edge in this row. Vanya opened one of the boxes and found that it was empty (marked in the figure with a shaded square). Find the fruits that are in the other boxes.

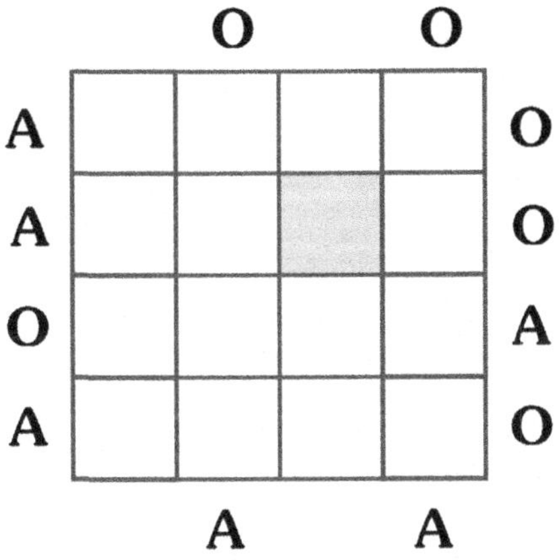

Problem 4. In 2015, Artyom will be 1 year older than the sum of the digits of his year of birth. In what year was Artyom born?

Problem 5. A figure made of matches is shown in the image below. You can see 5 equal squares. Move 4 matches so that only 3 squares can be seen. (There should be no extra matches)

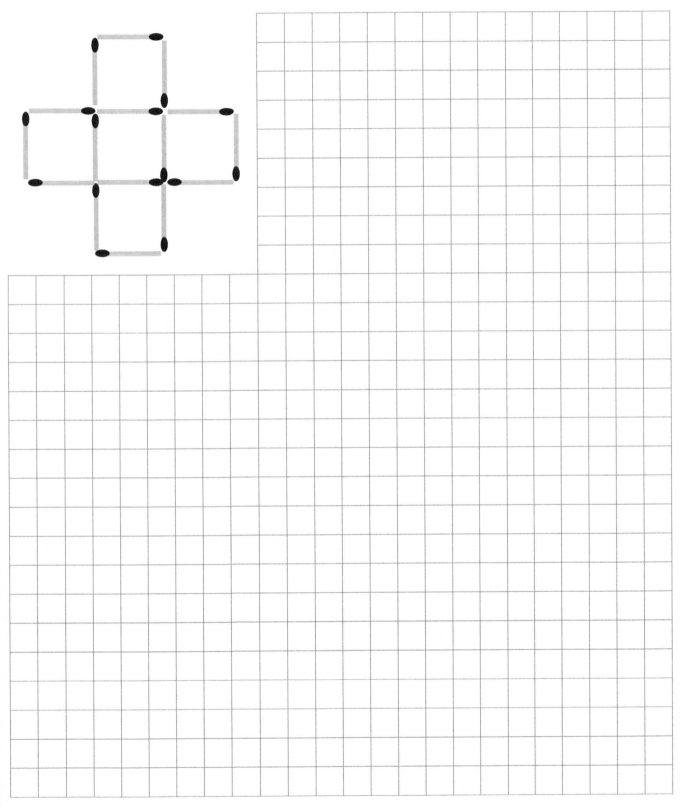

Problem 6. Some children observing a puppy noticed that if it barks, in a minute it eats; if it wags its tail, in a minute it plays; if it sneezes, in a minute it barks; if it eats, after a minute it wags its tail; if it plays, after a minute it sneezes. The puppy has sneezed now, what will it do in 12 minutes?

Problem 7. Krosh agreed with the Hedgehog to meet in the clearing. However, Krosh's watch is 15 minutes early, but he thinks it is 15 minutes late. And the Hedgehog's watch is 15 minutes late, but he thinks it is 15 minutes early. Who will come to the meeting first and how many minutes will he wait for the other?

Problem 8. Some children stayed in the classroom after the lessons. "Besides me, there are more boys than girls!" – said Nastya. "And besides me, there are more girls than boys!" – said Kolya. "They are both right!" – Misha said. What is the least number of boys and girls who stayed in the classroom?

Olympiad 2016

(XX Olympiad for Elementary School)

Problem 1. Find at least one solution to the numerical puzzle. If different letters correspond to different digits.

$$AAAA + BBB + AA + C = 2016$$

Problem 2. Santa Claus had 5 chocolates with different colored wrappers: red (R), yellow (Y), blue (B), orange (O) and green (G). Five children formed a circle and Santa Claus began to distribute the chocolates in a peculiar way, he skipped a child during the distribution (one yes and the other no). In what order were the chocolates distributed if the last one was in a yellow wrapper, and in the end everyone received chocolates as shown in the figure?

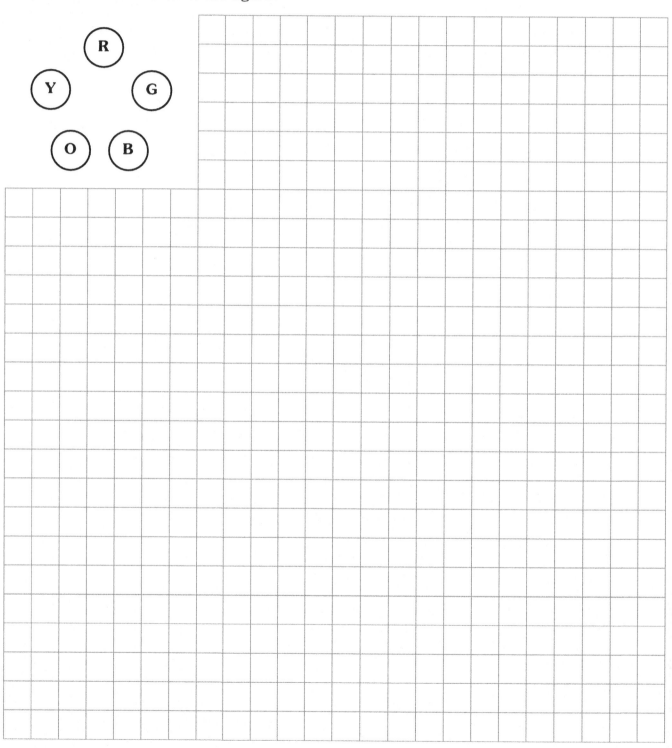

Problem 3. Place the numbers 1, 2, 3, 4 in the cells so that all 4 numbers are present in each row and in each column, and the indicated inequalities are satisfied.

Problem 4. Cut a 3 × 4 rectangle along the grid lines into two shapes of equal perimeter but unequal area.

Problem 5. Usually a child watches 5 cartoons on television before going to bed. But if the child is naughty during the day, he is forbidden to see some cartoons. For showing his tongue, the child Freken Bok misses cartoons # 1, 2 and 3. For not eating the meal - cartoons # 2, 4 and 5. For walking on the roof - cartoons # 1 and 5. For jumping over the bed - cartoons # 1 and 4. By playing with the cat - cartoon number 5. In the morning, Freken Bok decided that he wanted to see at least one cartoon today. How many of the listed antics can the child afford on this day?

Problem 6. Edward always says two statements, one of which is true and the other is not. He once said: "Yesterday was Wednesday. The day after tomorrow will be Tuesday". Then he thought a bit and said: "Today is Wednesday. Tuesday was the day before yesterday". What weekday is today?

Problem 7. A beetle is in the lower right cell of the board (see the figure). If it walked through four different cells, and in the fifth he decided to rest. Indicate which cell it could be in, if it is known that the beetle has made an equal number of turns to the left and to the right during its journey.

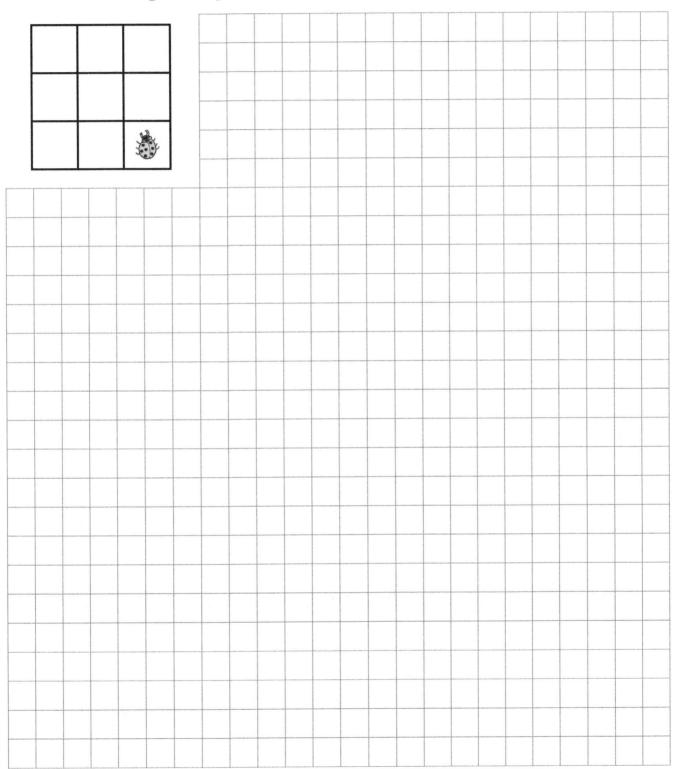

Problem 8. There are 4 coins in a row on the table. Two of them are known to be fake and weigh the same and are lighter than the real ones. It is also known that counterfeit coins are not neighbors. How to find both counterfeit coins in one weighing with a plate scale?

Olympiad 2017

(XXI Olympiad for Elementary School)

Problem 1. Place arithmetic signs to get the correct equality (you can use arithmetic signs and brackets as many times as you like):

$$1\ 2\ 0\ 2\ 2\ 0\ 1\ 7 = 5$$

Problem 2. Given a 7 × 7 squared board. Paint over the cells on this board so that no matter where on the board we choose a cross of five squares (see figure), it covers exactly one painted cell.

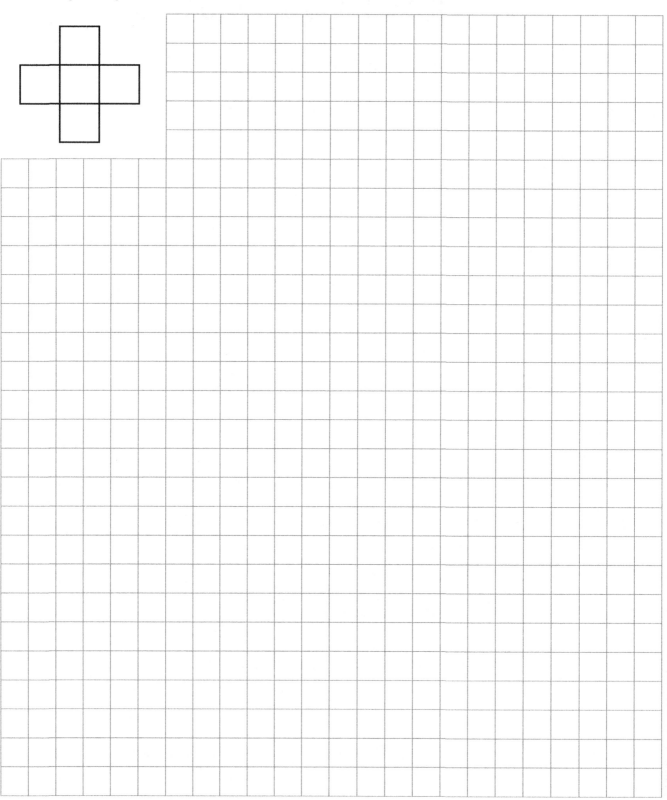

Problem 3. Tikhon represents the natural numbers of a single digit with the help of matches:

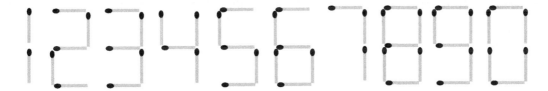

He represents the number "100" as shown in the figure below. Move 4 matches to get the largest possible number.

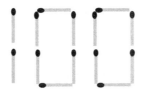

Problem 4. Grisha had three cats: Hasselblad, Vaska and Date. One of them had blue eyes, the other had yellow eyes, and the third had one yellow eye and the other green. If Date had the same eyes as Hasselblad, the total number of eyes of each color would be the same. What eye color does each cat have?

Problem 5. In what order did the pentomino figures fall from top to bottom in the game, if as a result they were positioned as shown in the figure?

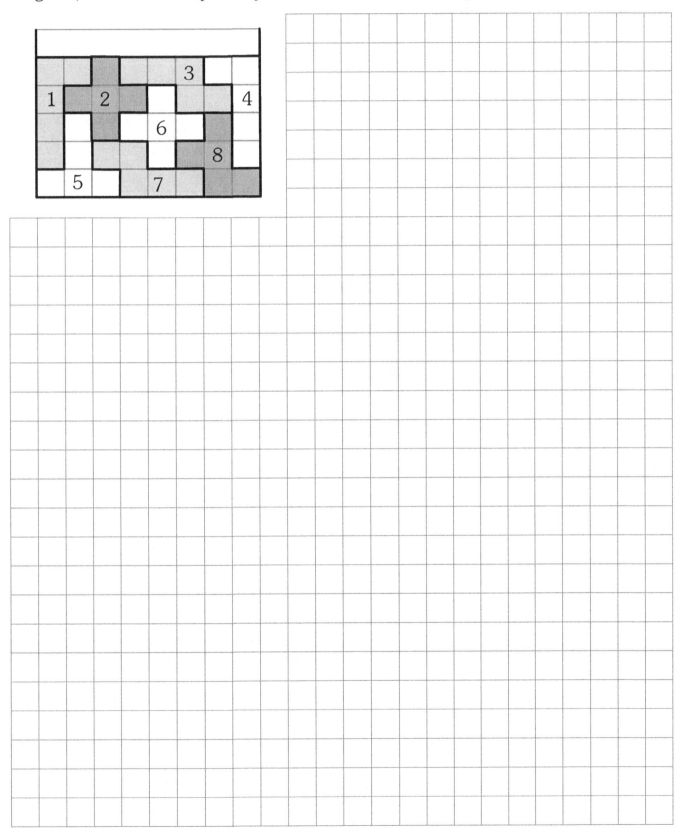

Problem 6. It was decided to number the seats in the single car of a Romashkov train, for which numbered cards were made. It turned out that there were 11 more cards with the number 1 than cards with the number 0. What is the least number of seats in this car?

Problem 7. Cut the Christmas tree in the figure with two straight cuts into multiple pieces so that all the pieces have the same number of balls.

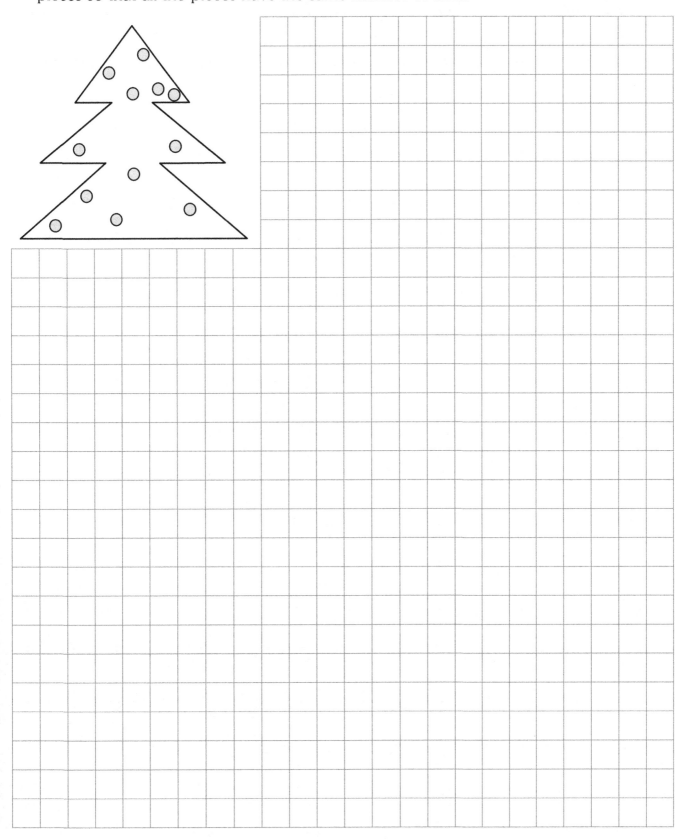

Problem 8. Petya counts the number of apartments in the building where he lives: 1,2,3, ... If the apartment number is divisible by 11, Petya sneezes, and if the floor number is divisible by 4, Petya coughs. The floor on which Petya first sneezed and coughed was the penultimate one. How many floors are there in the building where Petya lives, if there are 4 apartments on each floor of the building (including the first floor)?

Olympiad 2018

(XXII Olympiad for Elementary School)

Problem 1. Karabas-Barabas multiplied three different numbers greater than 1 and got 36. What numbers did Karabas-Barabas multiply?

Problem 2. A vending machine sells three types of chocolates A, B and C. Max wants to buy several chocolates so that he can make a 3×3 square with some of them (without breaking them). He notices that there are 1 type A, 3 type B and 7 type C chocolates in the machine. How much will it cost to perform this square with certainty, if each chocolate is worth 10 rubles? (The vending machine delivers the product randomly)

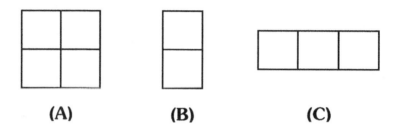

(A)   (B)   (C)

Problem 3. The owners of an art gallery decided to paint the hallway walls in 4 different colors so that the adjacent hallway walls have different colors. Show how they could do it. The gallery plan is shown in the figure below.

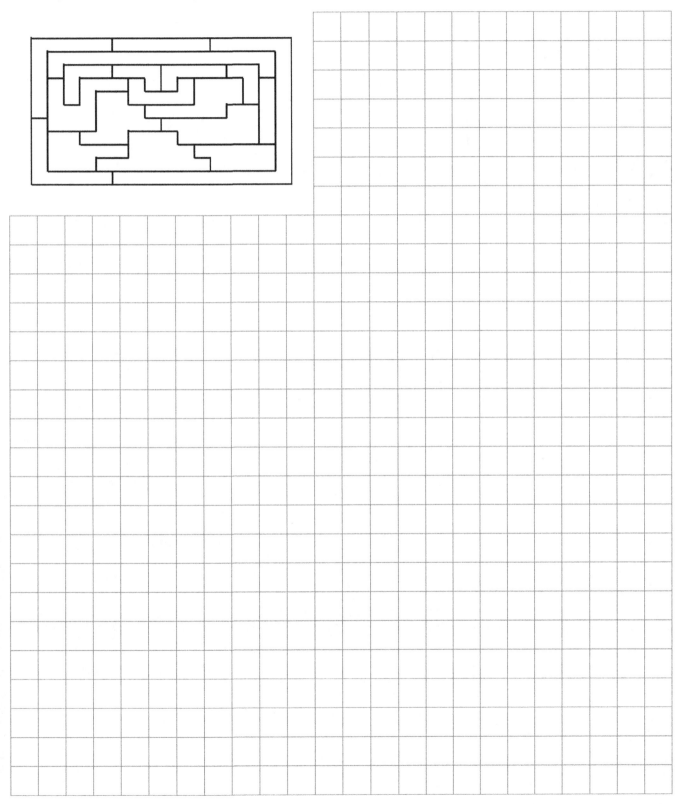

Problem 4. One day, Inga Borisovna's clock fell and shattered. The dial was divided into three pieces. Kolya realized that the sums of the numbers in each of these pieces are three consecutive numbers. Draw how the dial could have been broken.

Problem 5. Cut the shape along the grid lines into 4 equal parts. (The parts may be rotated)

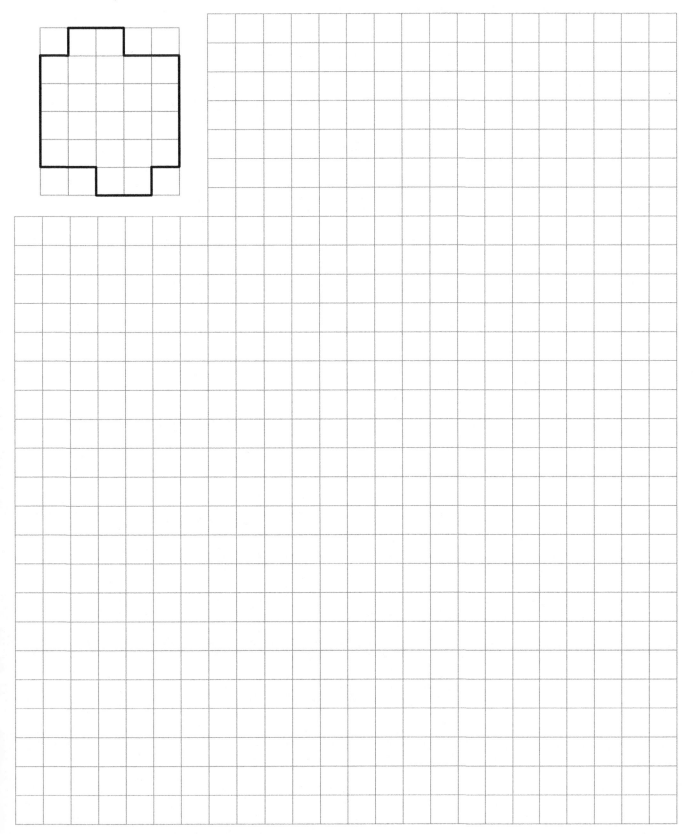

Problem 6. A swan, a crab and a fish try to move a boat together with the help of ropes for 2 hours. The swan pulls forward for 10 minutes, then 10 minutes backward, then 10 minutes to the left and 10 minutes to the right, again 10 minutes forward, and so on. The crab pulls 15 minutes backward, then 15 minutes to the left, then 15 minutes to the right, backward again, and so on. The fish pulls 20 minutes to the right, 20 minutes to the left, 20 minutes forward, again to the right, and so on. If the boat only moves when everyone is pulling in the same direction. How many minutes did the boat move during these 2 hours?

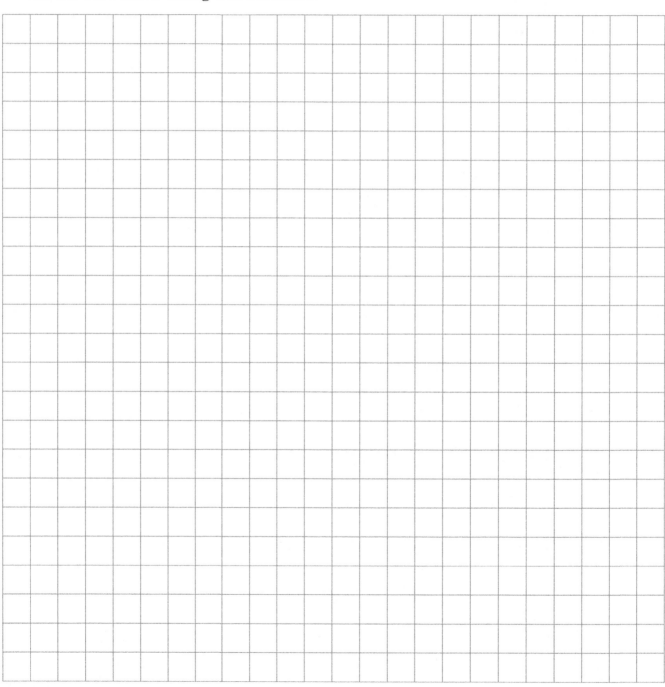

Problem 7. Dominoes with dots from 0 to 6 were placed in a spiral on a chess board (see figure). At one point, all the dominoes were placed. What are the cells that the last domino covered?

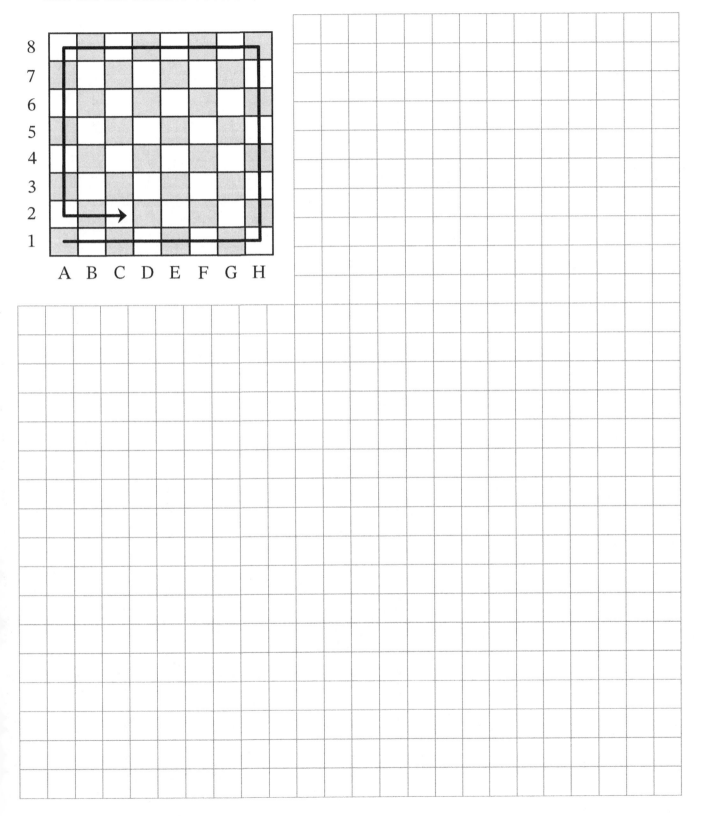

Problem 8. Three inhabitants of the island of gentlemen and liars met. One said, "We are all liars". The second objected: "We are all gentlemen!" and the third said nothing. Determine who is who if liars always lie and gentlemen always tell the truth.

Olympiad 2019

(XXIII Olympiad for Elementary School)

Problem 1. Carlson received a box of sweets. In the morning he ate a third of all sweets, at lunch he ate 2 less sweets than in the morning. And for dinner he finished the remaining 9 sweets. How many sweets were in the box?

Problem 2. Oleg folded a sheet of paper in four, as shown in the image, and made a straight cut. Then he unfolded the sheet. What figure could not turn out?

(A)   (B)   (C)   (D)   (E)

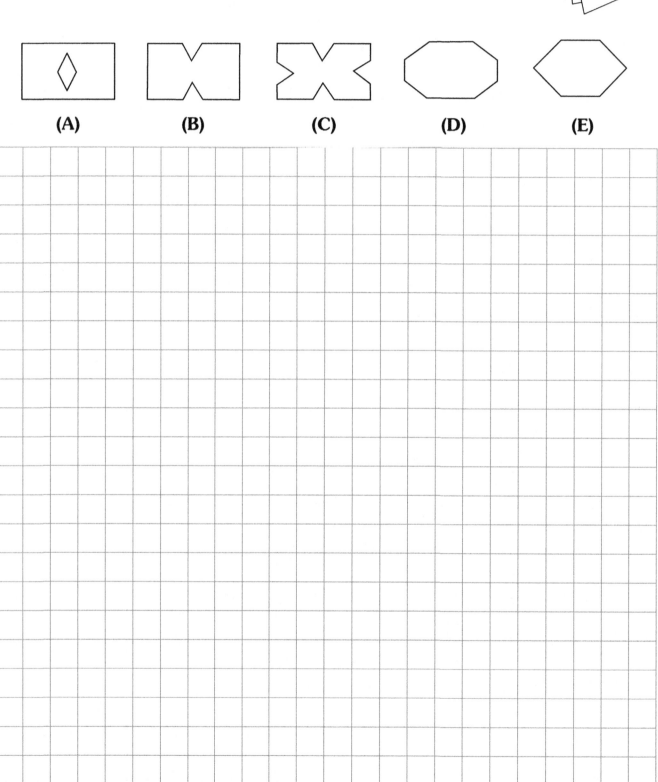

Problem 3. Write the numbers 1 through 9 in the small triangles so that in each large triangle the sum of the four numbers is 25.

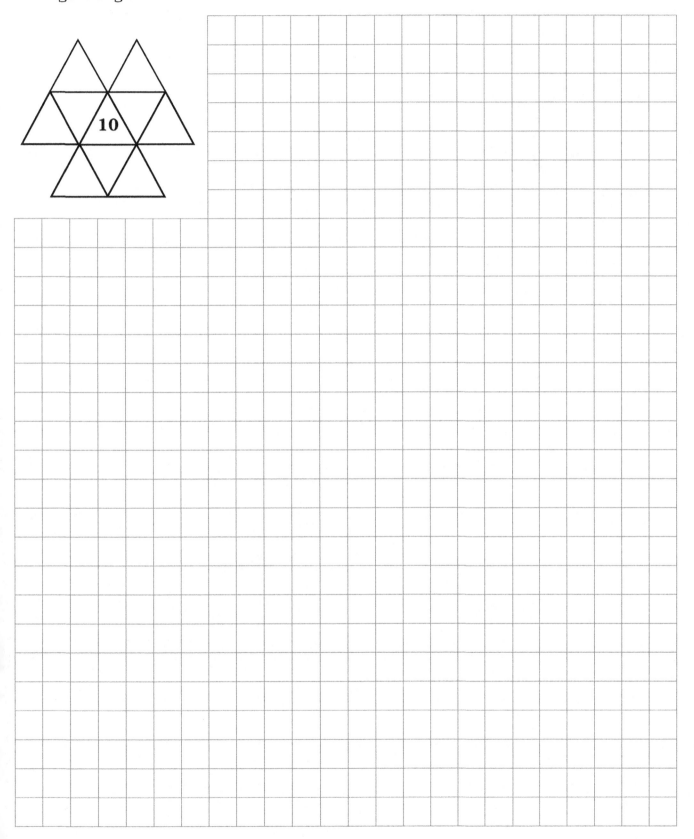

Problem 4. Luke Skywalker took possession of the magic monogram made of gold wire (see figure). To deprive Emperor Palpatine of power, Luke must cut the monogram into exactly 6 pieces with a laser cannon. Luke is tasked with shooting at just 3 points. How can you cope with this task?

Problem 5. Uncle Fyodor told Sharik a two-digit number. Sharik discovered that if this number is multiplied by 3, we get a two-digit number, and if we subtract 3 from the original number and then divide the result by 3, we will also get a two-digit number. What number did Uncle Fedor tell Sharik?

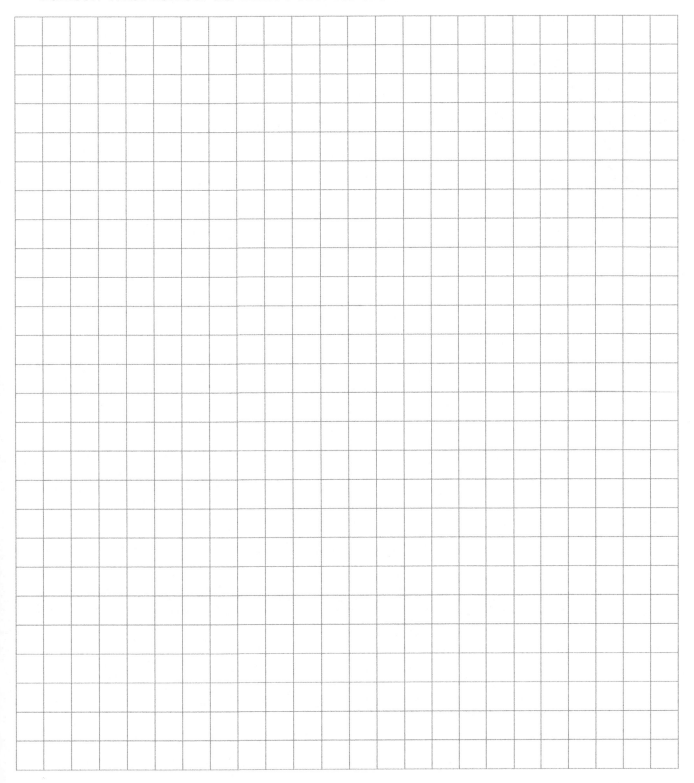

Problem 6. At Hogwarts, each of the faculties (Gryffindor, Slytherin, Ravenclaw, and Hufflepuff) has its own library and its own gym. The rules require that each faculty have its own path between the library and the gym, covered with a path of its own color, and the paths must not cross. Draw these 4 paths on the map shown below. (It can only move from room to room through its common side)

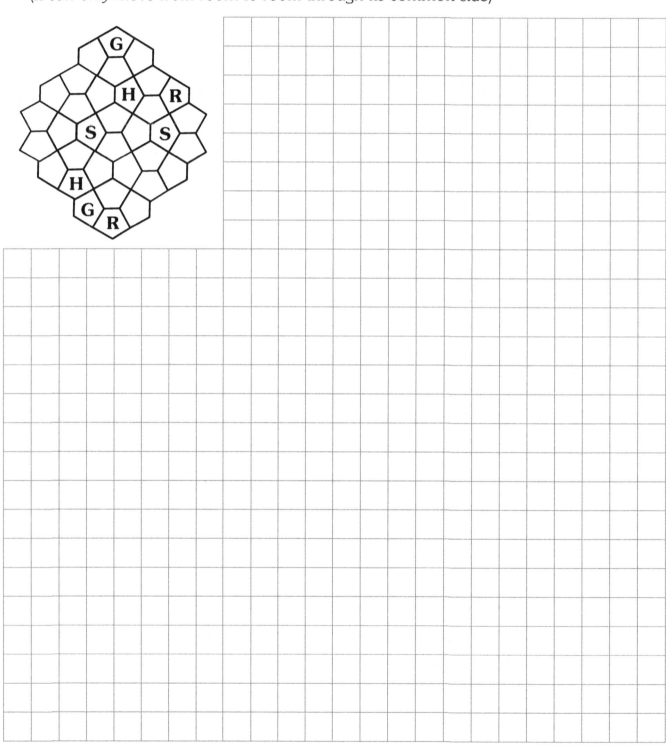

Problem 7. On April 12, *the shorties* launched a rocket to Mars. Znayka told her friends that the rocket would not be on Mars immediately, but after a while. And to the question "How long?" she silently showed one finger. Friends immediately shouted versions of what Znayka meant: second, minute, hour, day, week, month. To this Znayka replied: "One of you guessed right and the others were wrong, 24, 60, 168, 720, 3600 times". After what time, according to Znayka, will the rocket be on Mars?

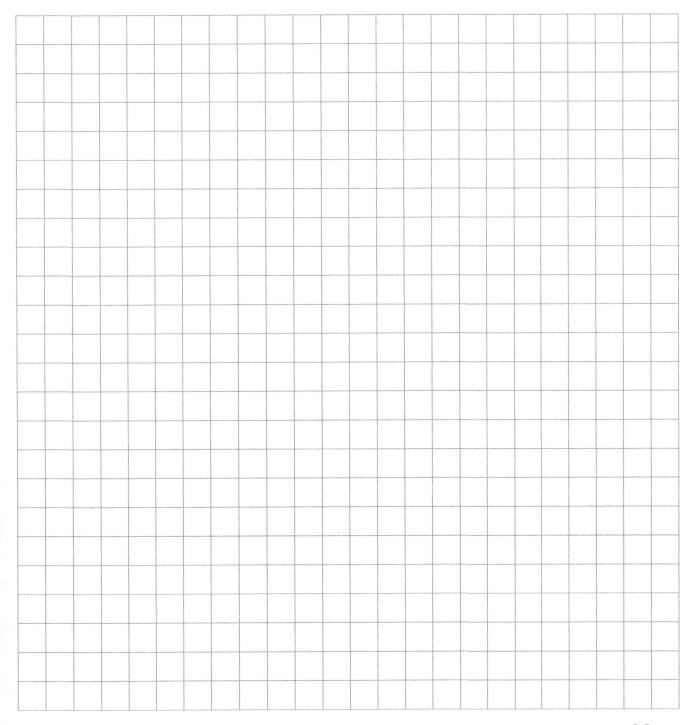

Problem 8. Three friends, Masha, Sveta and Dasha, were born the same year, but at different seasons of the year: winter, spring and summer. Sveta is younger than Dasha and more than six months have passed between Masha and Dasha's birthdays. When was each one born, if it is known that on September 1, they are not all the same number of years?

Olympiad 2020

(XXIV Olympiad for Elementary School)

Problem 1. Replace different letters with different digits so that you get the correct equality:

$$M + A + N + A + D + A = U \times P \times A$$

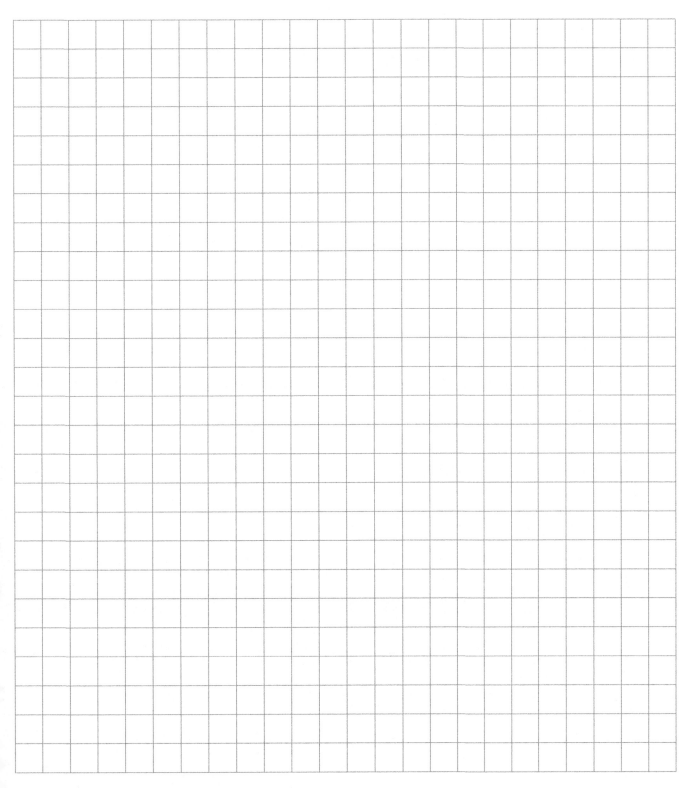

Problem 2. Masha (M), Petya (P) and Katya (K) live on the banks of a winding river. Alice got hold of a map of the area where the children live. Could you determine who lives on one bank and who lives on another? If yes, please indicate who lives on the same bank. If there are no other bodies of water in this area.

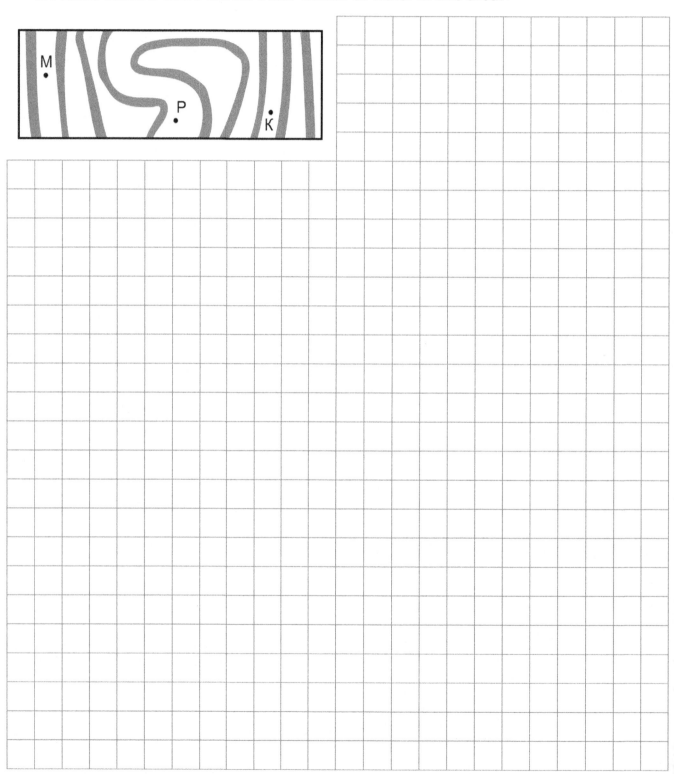

Problem 3. If you pour water into the structure in the figure to the side, vessel A will be filled first. Which vessel will fill first if you open the faucet in the structure in the figure below?

Problem 4. From home to school, Klim has two intersections with traffic lights showing green and red signals. Klim takes 2 minutes to walk from the first to the second traffic light. Klim knows that at each traffic light, the green and red lights are on for the same amount of time, 2 minutes each. It is known that it takes him 10 minutes to walk from home to the first traffic light and 10 minutes from the second traffic light to school. Once Klim left the house at 8:00 and saw that all the traffic lights were simultaneously turning green. What time will he get to school if he does not break the rules? (Klim crosses the street in 5 seconds)

Problem 5. If the smallest three-digit number that is not divisible by 2 is added to the largest three-digit number divisible by 2. What is the sum?

Problem 6. In a slot machine game, one coin must be placed at the same time in three circles connected in a triangle (One of which is placed in the central circle). Each circle shows the number of coins placed in it (in the central circle the screen is broken and the number is not visible). Petya started playing when the numbers in the circles were those in the left figure, and finished, when the numbers in the circles were those in the right figure. How many times did Petya play?

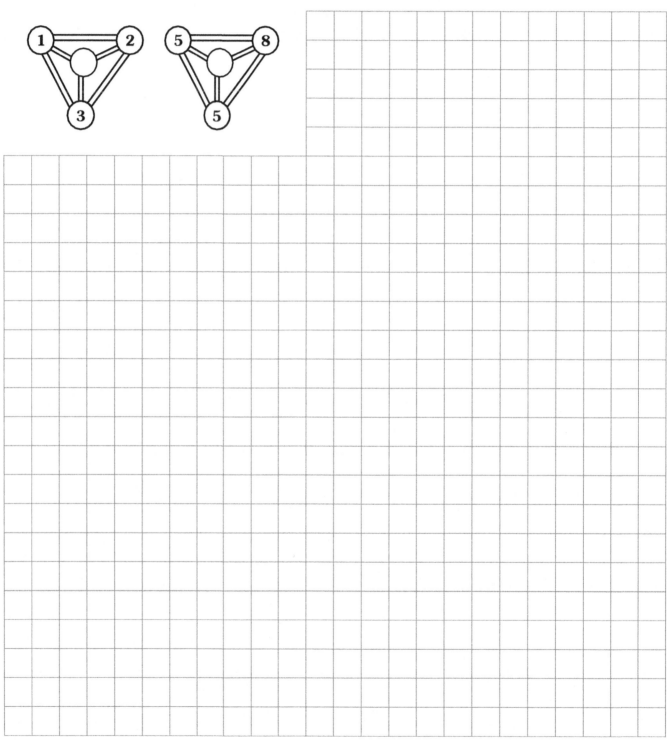

Problem 7. The Hogwarts Museum is divided into triangular rooms. A magic lantern installed in a room illuminates all rooms in three directions (as in the left figure). If a room is lit from three directions, it becomes invisible. Indicate all the invisible rooms in the plan shown in the right figure.

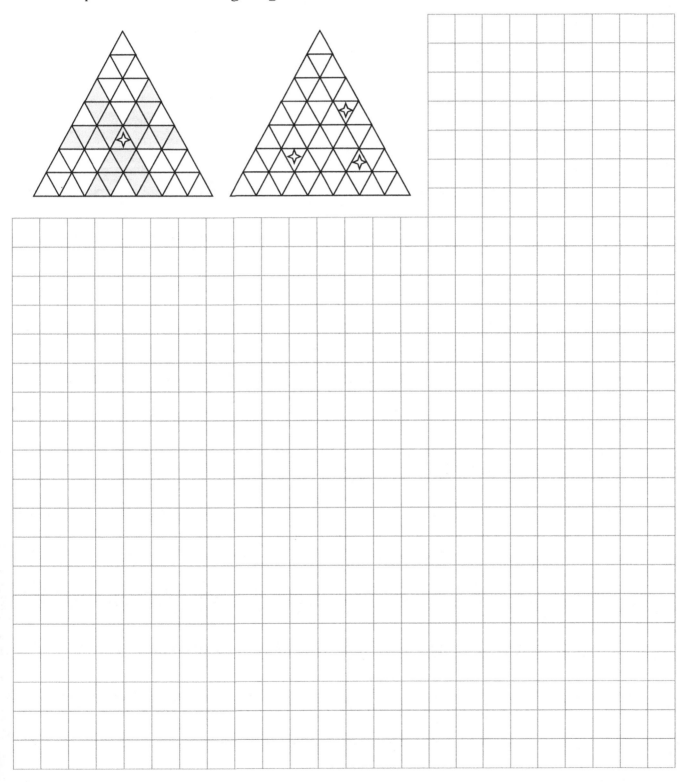

Problem 8. Three friends were playing dominoes. Each of them took a domino from the set and made three statements: "There are four points in my domino"; "My domino has an empty half"; "My domino has the same points in both halves". Which dominoes were taken from the set, if two statements are true and one is false?

Answers

Olympiad 2011

1. 4 people.

2. Yes, it is.

3. An example is shown in the figure:

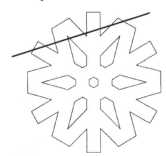

4. The heaviest is Hedgehog, the lightest is Nyusha.

5. 4 rings.

6. 1 hour.

7. (A) 11; (B) An example is shown in the figure below:

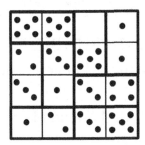

8. On the Rocky Island.

Olympiad 2012

1. 2000 = 2012 - 12.

2. The picture of apple number 3 is missing. The figure shows the matches for the rest of the apples.

6   4   1   5   2   7   8

3. The next figure shows an example.

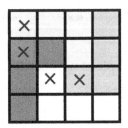

4. The filled table is shown below:

5	3	4	2	1
2	1	5	4	3
1	4	2	3	5
3	2	1	5	4
4	5	3	1	2

5. 9 footprints of Fyodor.

6. 3 boxes.

7. 9 Indians and 8 pale-faced.

Olympiad 2013

1. For example, 1765 + 46+ 202 = 2013 or 1674 + 57+ 282 = 2013.

2. My brother's name is Maxim.

3. 8 students.

4. An option is shown in the figure.

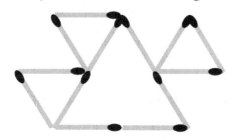

5. 4 times.

6. Vasya got piece number 2.

7. 20 minutes.

8. Gloria and Alex lied. Marty told the truth.

Olympiad 2014

1. Olya, Vasya, Petya, Masha.

2. 9 km.

3. Counterclockwise.

4. There are 3 more girls in the class.

5. The figure shows all possible options.

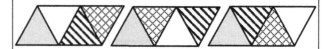

6. O+N+E+F+O+R+S+I+X = 35.

7. Popeye the sailor ate spinach for dinner last night.

8. Avoska was born on March 31.

Olympiad 2015

1. Six numbers. These are the numbers 111, 120, 102, 201, 210 and 300.

2. February 2nd. Since Kopatych got out only 10 times, by that time 80 days had passed. That means Kopatych got out of his den on the 80th. It will be on February 2nd.

3. The figure below shows the solution.

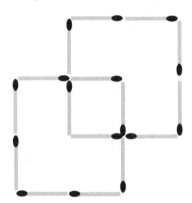

4. 2006 or 1988.

5. An example is shown as follows:

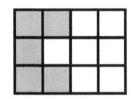

6. It will eat.

7. Krosh will arrive first and wait for Hedgehog for one hour (60 minutes).

8. 2 boys and 2 girls.

Olympiad 2016

1. 1111 + 888 + 11 + 6 = 2016.

2. ROGBY (if distributed counterclockwise) or ORBGY (if distributed clockwise).

3. An example is shown below:

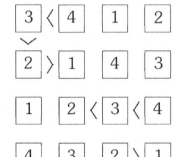

4. An option is shown in the figure:

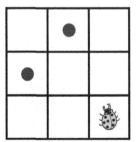

5. The child can afford 4 out of 5 antics.

6. Today is Thursday.

7. The two possible cells are marked with dots in the figure.

Olympiad 2017

1. An option is: (120: 2 - 20): (1 + 7) = 5

2. The figure below shows an example:

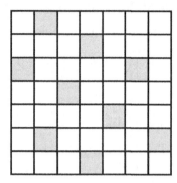

3. The number is 1 111 111

4. Vaska has blue eyes, Date has yellow eyes, and Hasselblad has multi-colored.

5. 5 - 7 - 8 - 6 - 2 - 3 - 4 - 1. For the last three figures there are also options 3 - 1 - 4 and 1 - 3 - 4.

6. 19 seats.

7. An example is shown in the figure.

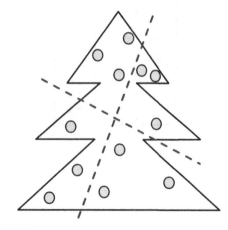

8. 21 floors.

Olympiad 2018

1. The numbers are 2, 3 and 6.

2. 50 rubles.

3. An option is shown in the figure.

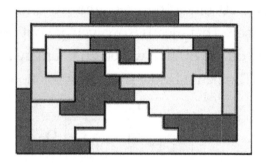

4. An example is shown below:

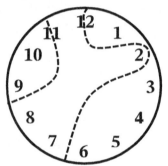

5. An example is as follows:

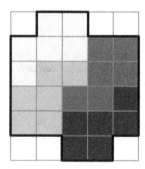

6. 15 minutes.

7. E6-F6.

8. The one who said nothing is a gentleman and the other two are liars.

Olympiad 2019

1. 21 sweets.

2. Option (C). In order to get figure C, we need to make two cuts. Below is how to obtain each of the figures.

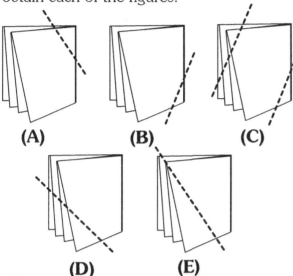

(A)   **(B)**   **(C)**

(D)   **(E)**

3. An example is shown in the figure below:

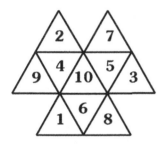

4. An option is shown in the figure below:

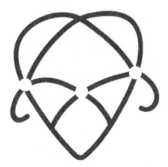

5. The number is 33.

If after multiplying by 3, the two-digit number is still two digits, then this number is not greater than 33. If after dividing by 3 the two-digit number is still two digits, then the number obtained by subtracting 3 from the original number is not less than 30, that is, the original number is not less than 33. Then, this number is at the same time not greater than 33 and not less than 33. Therefore, it can only be 33.

6. An option is shown in the figure:

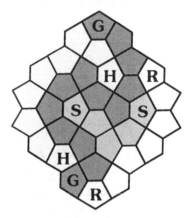

7. One hour later.

8. Dasha - in spring, Sveta - in summer, Masha - in winter, in December.

Olympiad 2020

1. For example, $6 + 1 + 5 + 1 + 0 + 1 = 2 \times 7 \times 1$.

2. Masha and Katya are on one bank; Roma is on the other.

3. Vessel 6 is filled first.

4. Klim will arrive at school at 8:26 and 5 seconds.

5. The sum is 1099.

6. Petya played 6 times.

7. In the figure, the invisible room is shaded gray.

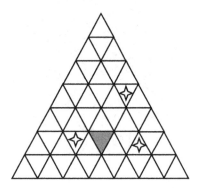

8. 0-0, 4-0 and 2-2. Note that two people cannot have the same true statements. Since two statements define the domino in a unique way. Therefore, three options must be considered and three types of dominoes obtained.

Made in the USA
Las Vegas, NV
13 February 2024